BRITAIN SINCE 1930

The Advance of Technology

Philip Sauvain

WAYLAND

BRITAIN SINCE 1930

The Advance of Technology
Leisure Time
Life at Home
Life at Work

Cover pictures: *Above: Demeter,* an Imperial Airways plane of the 1930s awaits passengers. *Below left:* In the 1990s, computers are a familiar part of life for many children. *Below right:* Eurostar trains travel from London to Paris through the Channel Tunnel.

Title page: The 1933 Morris 25.

Contents page: The London Telecom Tower.

Series editor: Francesca Motisi
Book editor: Liz Harman
Series designer: Joyce Chester
Production Controller: Carol Titchener

First published in 1996 by
Wayland (Publishers) Ltd
61, Western Road, Hove
East Sussex BN3 1JD, England

© Copyright 1996 Wayland (Publishers) Limited

British Library Cataloguing in Publication Data
Sauvain, Philip
    Advance of Technology. – (Britain Since 1930 Series)
    I. Title II. Series
    609.41

ISBN 0-7502-1653-0

Printed and bound by B.P.C. Paulton Books, Great Britain

Picture acknowledgements
The publishers would like to thank the following for allowing their photographs to be reproduced in this book: Hulton Deutsch 6 (top), 24 (top); Impact/Piers Cavendish *cover* (bottom right), 26 (bottom), Impact/Homer Sykes 27 (top), Impact/David Reed 29 (middle); The Robert Opie Collection 10 (right); Philip Sauvain *title and contents pages*, 4, 5, 7, 8, 9, 10 (top), 11, 12, 13, 14, 15, 16, 17, 18, 20, 21, 22, 23, 24 (bottom), 25 (top), 26 (top), 27 (bottom), 28; Topham Picturepoint *cover* (top), 6 (bottom), 28 (bottom), 29 (top); Wayland Picture Library *cover* (bottom left), 19.

Contents

In the home

Silvry age and ruddy youth
Bear witness to this golden truth
"SUNLIGHT IS BEST"

⇐ Early this century, washing clothes by hand took all day and filled the kitchen with steam. Many homes in the 1930s were still without the mains services, such as gas, water and electricity, which most of us take for granted today.

There have been many striking advances in technology in the last sixty or seventy years. Motoring, aviation, telephones, radio, television and computers have all had a big effect on people's lives, while electricity has made possible the use of many labour-saving devices as well as providing light, heat and power.

Heat and light before electricity

'Before electricity came along the fire was our only way to heat and boil kettles and for cooking. For lighting we had to carry a candle to bed. The cheer that went up the day we had electricity was heart-warming.'

An advert for candle lamps in a department-store catalogue for 1933–34. Many farmhouses and homes in remote parts of the country were without electricity until the 1950s and 1960s. ⇓

4

Many homes, especially those in rural areas, first received electricity in the early 1930s. Magazines and newspapers published at the time described the effect that electrification was having on the country.

An article about electricity in *The Daily Telegraph*, January 1932

'Within three months the electricity "grid" will be completed in East Anglia. Already in this predominantly rural area, village lanes are now brightly lit by electricity. Electric lamps, cookers, irons, washing machines, and vacuum cleaners are replacing coal fires, oil lamps and hard manual labour in farms and cottages. Cows are being milked, milk separated, corn threshed, hay cut, and turnips mashed, all by electricity.'

New electric pylons in 1934. ⇨

⇐ A demonstration of a washing machine at the London School of Electrical Domestic Science in 1934.

⇑ A woman trying out her new vacuum cleaner in 1951.

There was widespread use of portable electric heaters and gas- and oil-fired central heating from the 1960s onwards. This made it unnecessary to light coal fires in every room, so there was less dirt. Fitted carpets and vacuum cleaners also helped to make houses easier to clean. Since less work was needed to prepare food and keep a home and its contents clean, people had more leisure time. Fewer servants were needed in big houses.

Improvements to drains and water supply in the 1930s and again in the 50s and 60s also helped to improve living conditions for millions of people.

A woman remembers houses in the country

'I was born in a cottage with no electricity or water. When I was four years old we moved to a brand new council house in the village, but we still had no electricity and had to fetch the water from an outside pump. In 1958 I got married and in 1959 we were allocated a council house in Kedington. We had electric light, of course, but still no hot water or bathroom and the toilet was outside.'

Since the 1950s, many other advances in modern technology have helped to reduce housework. Duvets instead of blankets made it quicker to make a bed. Freezers and fridges replaced the old-fashioned larder. Electric ovens, toasters, coffee machines, kettles, carving knives, tin-openers, microwave ovens, deep-fat fryers and other electrical appliances made it easier to prepare meals. Dishwashers, automatic washing machines, tumble driers and steam irons made light work of cleaning chores.

Household appliances in 1932 – gas cooker 'to meet the growing demand for a modern and economical cooker'; radiator 'for use with water or steam'; chemical closet (lavatory) 'used where there is no drainage system'; enamel bath 'to meet the demand for baths for small properties'. ⇩

Motoring

It was not until the 1920s that the first cheap motor cars were made in Britain. Before then, only the very rich had been able to afford the hand-made luxury cars produced by manufacturers like Daimler and Armstrong Siddeley. Cheap cars in the 1930s made it possible for many more people to own a car, although it was not yet within the reach of the average family.

First ride in a motor car in the 1930s

'I can only just remember my first trip in a car. It was 1938 and I was five years old. We travelled from Coventry – seven adults and two children plus heaps of luggage – in two small cars. We left Coventry at 7.15 am and travelled south to Pevensey (near Eastbourne) for a fortnight's seaside holiday. According to my grandfather's diary, it took him eleven hours to drive the 166 miles [267 km] to the sea – at an average speed of 15 mph [24 kph].'

⇧ An advert for a luxury car in 1933 – the Morris 25.

An advert for the 1938 Ford V8 car. At the time, it cost £240; more than some houses. ⇩

The growth in the number of cars on the road was halted in 1939 with the outbreak of the Second World War. After the war had ended in 1945, it took a number of years for the motor industry to recover and people had to wait years for their new cars to be delivered. Rising living standards in the 1950s increased the demand for new cars. Since then, many more people have been able to afford to buy their own vehicle – as you can see from the table on the right.

Year	Number of private cars
1930	1 million
1939	2 million
1957	4 million
1964	8 million
1984	16 million
1992	22 million

The receipt for a new car, sold to a Devonshire doctor in 1937. The car was delivered by train, rather than by road, on a car transporter. ⇩

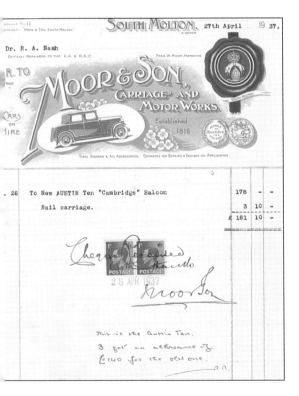

In a class of its own for "Quality First" Features

The "Quality First" MORRIS

Take a searching look at the next modern Morris you see. In its styling, interior appointments, superfine finish and in its performance too, it has entered a new and higher class in economical motoring. Until that happy day comes when you take delivery of a new Morris keep your present model in first-class condition by availing yourself of Morris Quality First Service. There are over 2,000 authorised dealers throughout Britain who are specialists in Morris methods and carry stocks of factory-inspected spares.

An advert for the 1951 Morris Oxford. ⇨

⇐ The Ford Popular in 1958. Many more people could afford to buy cars like this than before the Second World War.

The new Austin Seven in 1959. Mini cars like this were very different from previous designs. Putting the engine sideways provided extra space. This was used to make the car body much smaller than most other cars, making it easier to park. Yet it could be just as fast and used much less petrol. ⇩

A first car in 1958

'My first car was a Ford Popular. It had a long narrow window at the back like the slit in a letterbox. There was no heater and the windscreen wipers only worked when the car was moving. It had three forward gears and you had to press the clutch pedal twice each time you changed [gear] up or down. Top speed was about 50 mph [80 kph]. It was probably the most reliable car I have ever had. There was so little to go wrong!'

As you can see from the pictures, the shape and design of motor cars changed over the years. New advances in technology made it possible to make cars faster, more comfortable, more economical to run – and much safer.

Car designers tried to do several things at once – keep costs low but make their vehicles as fast and as spacious as possible. This is why the development of the Austin Seven and Morris Mini-Minor caused such a stir when they were introduced in 1959.

The new AUSTIN SE7EN

The rapid increase in the number of motor vehicles since the 1920s and 1930s has meant many changes to Britain's roads, especially in towns and cities. These photographs of Piccadilly Circus in London were taken at intervals of forty years and show how methods of traffic control have changed.

⇦ Chaos at Piccadilly Circus in about 1915. Traffic approaches the Circus from all directions. Imagine what it was like to cross the road on foot!

Piccadilly Circus in about 1955. Traffic controls have been introduced. Now, vehicles move clockwise, in an orderly fashion, around the statue of Eros in the middle. ⇨

⇦ Piccadilly Circus in 1995. The area around the statue of Eros has been pedestrianized (turned into a small area for pedestrians only). One-way traffic has been introduced on the roads leading to the Circus and the flow of vehicles is controlled by traffic lights operated by computers.

⇦ A Belisha beacon crossing and traffic lights, soon after their introduction in the early 1930s.

Road safety landmarks	
1930	All cars have to be insured in case of an accident.
1931	Traffic lights in general use.
1934	Catseyes installed to show up the middle of the road at night. Pedestrian crossings marked out by yellow Belisha Beacons.
1935	30 mph (48 kph) speed limit in towns. Compulsory driving tests for new motorists.
1965	Maximum speed limit of 70 mph (112 kph) on motorways and dual carriageways.
1983	Compulsory for drivers and front-seat passengers to wear seat belts
1991	Wearing of rear seat belts made compulsory for adults.

Largely as a result of the safety measures shown in the table above, road deaths fell from 7,000 a year in 1930 to 4,200 a year in 1992, even though there were sixteen times as many vehicles on the roads.

Motoring changed people's lives in many other ways. It helped people to see more of their own country, and made it easier for people to change jobs and move to other parts of Britain in search of work. It also changed the way in which many people spent their leisure time.

Although Germany and Italy both enjoyed the benefit of fast new motorways in the 1930s, motorists in Britain had to wait until 1958, when part of the M6 – the Preston Bypass – was opened in Lancashire. Since then, nearly 3,200 kilometress of motorway have been built, drastically cutting the time needed to make a long journey. Drawbacks of motorways include the destruction of farms and fields to make way for new roads and polluting the atmosphere with fumes from the extra traffic attracted to these roads.

We often think of caravanning as modern. But, as you can see from this cartoon published in *Punch* magazine in 1933, motorists were already towing caravans into the country or to the coast. ⇩

⇦ Building new bridges to carry motorways and dual carriageways has also altered the look of the countryside. The Forth Road Bridge, shown here, was opened in 1964. It linked Edinburgh and the M8 to the south of the Forth Estuary with the M90 to the north.

Aviation

In 1930, aviation was still quite new. Just twenty-seven years earlier, in 1903, Orville Wright had become the first human being to fly. Some people thought the future for aviation lay with airships (known as dirigibles), which could carry more passengers than early aeroplanes. An airship was a huge balloon filled with hydrogen gas, which floated on air and was powered by an engine. However, hydrogen gas can be highly explosive, and some airships were involved in terrible accidents, like the one described below. Most people thought that airships were more dangerous than airliners.

⇧ An airship moored to a mast in about 1934.

GREAT AIRSHIP STRIKES A HILL AFTER BATTLE WITH A STORM

SLEEPING PASSENGERS ENVELOPED BY SWIFTLY RUSHING FLAMES

The giant airship R101, which left Cardington [near Bedford] at 7 pm on Saturday for India, crashed near Beauvais, France, at 2.05 am yesterday, and forty-six of those on board were burned to death. [There were eight survivors.]

Low-lying clouds had prevented the R101 rising, and the storm, which left masses of rainwater on the top of the airship, forced her down until she struck a low hill. The shock broke the R101 and the dangerous hydrogen gas inside exploded. The airship was soon burnt out.

⇦ Report in the *Daily Express*, 6 October 1930.

By contrast, the aeroplanes used by the British airline Imperial Airways had never lost a passenger. In 1930 its airliners were already flying passengers to India. The airline started new services to Central Africa in 1931, to Capetown, South Africa in 1932 and to Singapore in 1933. Two years later it began to fly passengers from London to Brisbane in Australia. The journey took ten days but even that was much quicker than sailing there by boat. London's main airports at that time were at Heston and Croydon. Imperial Airways flew airliners like the one on the right. In 1939, Imperial Airways merged with another airline, British Airways, to form the British Overseas Airways Corporation (BOAC).

The following description was given by an Imperial Airways airliner passenger in 1938:

⇧ Imperial Airways' Handley Page airliner *Horatius* at Le Bourget Airport, Paris, France in 1937. This massive biplane carried thirty-eight passengers and a crew of four. It travelled at an average speed of 160 kph.

The comfort of an airliner

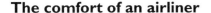

'It offers such amenities as today's newspapers, unrestricted leg-space and roomy lounge seats with comfortable cushions. Walls and doors are panelled in wood. Silk-curtained windows, shaded reading lamps, carpeted floors and a staff of stewards combine to make the journey no less comfortable than one by sea or by land.'

⇦ A German Junkers G38 airliner landing at London's Croydon Airport in about 1938. The Junkers was a monoplane, had a maximum speed of 210 kph and carried thirty-four passengers and a crew of seven.

Catching the 1.30 pm flight from London to Brussels, Belgium in 1938

'The passenger reaches Airways House, Victoria Station, at 12.35 pm. Baggage is surrendered on entering the hall. At the counter the passenger hands in his passport and stands on the scales. A few moments later the coach sets off for the forty-minutes' drive to Croydon Airport.

The passengers enter the huge forty-two seater by a covered gangway. Uniformed stewards show them to seats. At a signal from the control officer, the machine turns into the wind and roars across the aerodrome. In the centre of the ground, in enormous white letters, is the name CROYDON. The wheels leave the ground at the letter D. A few moment's later the machine's shadow is seen on the grass far below.'

'Tables are laid for luncheon. Today's menu is typical of modern airline catering: Sole; Veal; Roast Beef; Roast Chicken; York Ham; Sherry Trifle; Fruit Salad; Cheese; Coffee. Before the second course has been served, we have left the English coast and there is barely time to take coffee before we touch down in Brussels [at about 3.30 pm]. For convenience, luxury and speed, modern air travel is hard to beat.'

Surprisingly, the first regular British airline service from London to New York, USA, did not begin until 1946. Even then, it took twenty hours to cross the Atlantic.

B.O.A.C. takes good care of you

Words can't add much to a picture like this

People enjoy themselves like this every evening in the lower-deck lounge on B.O.A.C. "Monarch" luxury services across the Atlantic. Can you imagine anything more inviting, more comfortable, more attractive? And this lower-deck lounge is only one of the pleasant surprises in store for you. You'll appreciate and never forget the double-decked *Stratocruiser* spaciousness . . . the delicious complimentary meals and mealtime drinks . . . the attentive night-long service. All climaxed by deep, undisturbed sleep in a foam-soft berth at slight extra cost. You wake up refreshed — in time for breakfast in bed, if you wish!

LUXURY
Monarch
SERVICES

Overnight from London
to New York or Montreal
direct—no extra fare!

⇦ A BOAC advert in *The Sphere* magazine promotes its luxury trans-Atlantic Monarch services in 1953. Overnight passengers could even order breakfast in bed.

Meanwhile, the Second World War had sparked off a number of startling changes in air travel. The invention of the jet engine for use in fighter planes led to the development of the world's first jet airliner in 1949. This was the British-made De Havilland *Comet*. It led the way in making affordable, fast air travel possible and was soon followed by many other jet airliners around the world. Air travel increased rapidly. London's new Heathrow Airport, built just after the end of the war, soon became the busiest international airport in the world.

⇧ The world's first jet airliner *Comet* featured on the front cover of *The Aeroplane* magazine in September 1955.

The outstanding achievement of the *Comet* was followed in 1969 by the supersonic, Anglo-French airliner *Concorde* which flew for the first time on 2 March 1969. *Concorde* could carry over 100 passengers at more than 2300 kph – three times as many passengers as the Junkers airliner on page 16 and at ten times the speed!

Cheap air fares and rapid travel made it possible for people to fly on business to a foreign city and return home the same day. Even day trips to New York were possible using *Concorde*. People travelling to Australia in the 1990s can fly by airliner in twenty-four hours instead of taking a sea trip of many weeks. Rock bands and opera stars appear in London one day and in New York the next. Young people think nothing of taking holidays to Thailand or Australia. Their parents, at the same age, counted themselves lucky to go to Paris or Spain!

⇧ *Concorde* taking off.

Broadcasting and telecommunications

The telephone made it possible for people to speak to each other anywhere in the world. Seventy years ago, there were one million telephones in Britain. Although the number had increased to three million by 1939, most people still thought of the telephone as a luxury. As living standards rose in the 1950s and 1960s, many more people had telephones installed. By 1995 there were thirty million of them, including three million mobile telephones. Today, climbers even take mobile telephones with them to the summit of Ben Nevis, Scotland, in case they get lost!

Most telephone users had to wait until the 1930s before being able to dial local numbers direct. Until then, all calls were connected by an operator at the telephone exchange. Long-distance calls could not be dialled direct until 1958. ⇩

The first regular radio broadcasts in Britain began in the early 1920s with the founding of the British Broadcasting Corporation (BBC)

Radio was an instant success. A cartoon in *Punch* magazine in 1929 showed town roofs bristling with radio aerials. People even made their radios from kits, since they were much cheaper than those in the shops.

Broadcasting House in Central London was opened in 1932. ⇩

⇧ The London Telecom Tower was built in 1964 to send and receive telephone calls (and also radio and television signals). It beams signals from the dish-type aerials you can see on the tower.

The TELSEN RADIOMAG
Volume I
Price Threepence
Number 2
How to Buil
THE TELSEN SHORTWAVE
THE TELSEN S.W. ADAPTO
THE TELSEN TRIPLE THRE
THE TELSEN SONGSTER TW
THE TELSEN CONQUEROR THRE
THE TELSEN COMMODORE THRE
THE TELSEN EMPIRE FOU

⇦ The front cover of a 'do-it-yourself' radio magazine in 1932. The instructions inside are not quite as simple as the cover suggests!

In 1937, this luxury radio set cost £26 – three months' wages for many workers. ⇩

A G.E.C. PRODUCT
G.E.C.
FIDELITY ALL-WAVE 8

Listening to the radio became a national pastime. People stayed in at night to hear their favourite programmes instead of going out. By 1939, there were over eleven million radio sets in Britain. During the Second World War, radio provided entertainment while supplying people with news of the war.

Radio in wartime

'I don't think I really appreciated the wonder that was radio. I just took it for granted that by turning a knob I could listen to the war news one minute and a comedy show the next.'

The popularity of radio reached its peak during the war but, by the late 1950s, it was losing listeners rapidly to a new form of entertainment – watching television.

The scientist John Logie Baird first experimented with television in 1927. He found it difficult to convince the experts, as a writer called Sydney Moseley noted in his diary on 1 August 1928:

> *'Saw television! Met John Logie Baird. We sat and chatted. He told me he is having a bad time with the scoffers and sceptics – including the BBC and part of the technical press – who are trying to ridicule and kill his invention of television at birth.'*

⇦ John Logie Baird watches the test transmission of the first television play on 14 July 1930.

Baird's television system worked, but the pictures were not as good as those of a rival electronic system developed by EMI-Marconi in the 1930s. After testing out both systems, the BBC eventually used the EMI-Marconi system soon after the world's first television service was opened on 2 November 1936 at Alexandra Palace, London. Six months later, television showed what it could really do when the BBC televised the Coronation procession of King George VI in May 1937.

Watching the coronation on television in 1937

'Horse and foot, the Coronation procession marched into London homes on the television screen yesterday – seen instantaneously by 30,000 watchers far from the Royal route in drawing rooms, shops, offices and cinemas. Televiewers, some of whom were watching miles away at Brighton, had been spared the long wait on the procession route and the packed discomfort of the pavements.'

⇧ An early television set, which went on show at an exhibition in 1939.

King George VI's coronation procession to Westminster Abbey, in 1937, which was broadcast on television in black and white. ⇩

Two years later, television programmes went off the air during the Second World War. They began again in June 1946, but made little impact until the coronation of Queen Elizabeth II was televised in 1953. Thousands of people bought or rented sets for the occasion. In the ten years from 1949 to 1959, the number of television sets in Britain increased from about 125,000 to well over nine million. This was partly due to the opening of a second channel – ITV – in 1955.

There have been many exciting new developments in television. The launching of the satellite transmitter *Telstar* in 1962 made it possible to see live pictures from the USA on British television. A third channel – BBC 2 – was opened in 1964, and the first colour television broadcasts began in 1967. The most outstanding event in television history took place in 1969:

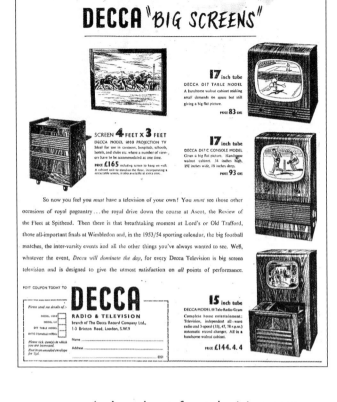

⇧ An advert for television sets at the time of the coronation in June 1953.

Watching the Moon landing, 21 July 1969

'I got up about 4 am, crept downstairs and switched on the television set. I couldn't believe my eyes. Armstrong and Aldrin were getting out of the lunar module and walking on the surface of the moon. The words on screen – "Live from the Moon" – made it all the more unbelievable. That they should be standing there, a quarter of a million miles [360,000] away, was remarkable enough. But that I could sit at home and watch them do so was even more of a miracle.'

Television enabled viewers to watch live coverage of the moon landing in 1969. ⇨

Changes in industry, medicine and science

Since 1930, advances in technology have changed many jobs in industry, farming, offices and transport. Using automatic methods to control machines – called automation – has meant that fewer workers are needed.

Industries have changed in many other ways as well. They used to use steam, produced by burning coal or coke in a boiler, to power their machines. Now they use fuels such as gas, electricity and oil, which are cleaner, quieter and easier to use.

Today, some automated factory processes are done by robots which can perform a number of intricate tasks accurately, such as spot welding on a motor car. ⇩

⇧ A guidebook for the Festival of Britain in 1951. This exhibition, held on London's South Bank, was to show off what was best in British industry and technology.

⇦ The discovery of natural gas and crude oil under the bed of the North Sea in the 1960s meant that Britain could produce all the oil it required instead of buying it from abroad.

The Docklands Light Railway caused a stir when it was opened in 1987. Instead of having a driver, the trains are controlled by a computer, which picks up information about the line ahead at each station on its route. ⇩

Since the 1950s, the use of computers and calculators, has had a big effect on transport, factories and businesses, such as offices and supermarkets. The time they can save has meant that many workers have lost their jobs as a result of these changes.

The first calculators

'I bought a mains electronic calculator for my business in about 1973. It seemed like a miracle at the time – being able to press two or three keys and flash up the answer to a complicated sum. It cost us £95 – worth well over £500 today – and yet it could only add, subtract, multiply and divide. Nowadays you can buy a pocket calculator which does all that for less than £5!'

Advances in technology have also changed the appearance of Britain's towns and cities. London now has several skyscrapers. Forty years ago the tallest buildings in London were its cathedrals and churches.

⇦ The City of London in 1928, seen from the tower of St Bride's Church near Fleet Street.

Advances in modern technology have also changed the way doctors and hospitals work. New machines and new techniques have made it possible to save lives by transplant surgery. Electronic scanners and miniature television cameras take pictures inside the human body. Laser beams are used to perform delicate operations on the eye.

Having tonsils removed in 1938

'I was five years old and remember waking up on a huge mattress spitting blood. There were several other children on the mattress as well doing the same – most of them wailing. We went home the same day on the bus!

Someone else I know had her tonsils removed by a local doctor – lying on the kitchen table at home in a farmhouse with no electricity, no indoor lavatory and no bathroom!'

⇧ Experts used to think the clay under London could not support the weight of a very high building. However, advances in technology have made it possible to build skyscrapers such as the Nat West Tower (seen here) and Canary Wharf (which can be seen in the photograph on page 27).

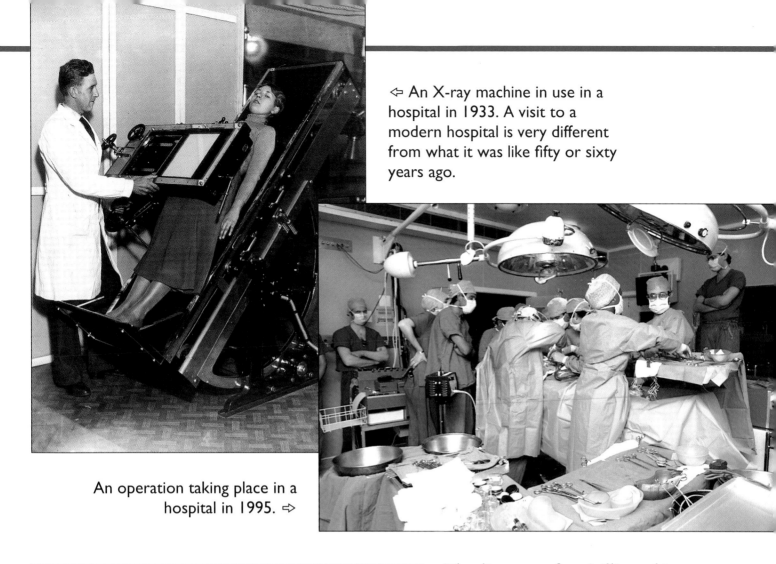

⇐ An X-ray machine in use in a hospital in 1933. A visit to a modern hospital is very different from what it was like fifty or sixty years ago.

An operation taking place in a hospital in 1995. ⇨

The discovery of penicillin and its development and use during the Second World War changed the way in which doctors treated infections. Since then, other antibiotics have been used to speed up the recovery of people suffering from injuries or diseases.

Science has also helped farmers to harvest more crops from the land. They use pesticides to kill insects and fertilizers to enrich the soil. Machines such as tractors and combine harvesters make farm work much easier.

⇐ The automatic milking machine made milking much easier and fewer farm workers were needed.

Glossary

Aerodrome A small airport or airfield.

Airships Huge gas-filled balloons with compartments underneath for passengers and crew. They were flown using propellers driven by an oil-fired engine.

Antibiotics Substances used to treat or prevent infections.

Automation The use of machinery that works on its own without the constant attention of a worker.

Aviation Travel by air.

Belisha beacon crossing A pedestrian crossing marked by yellow globes mounted on black and white poles. Named after Leslie Hore-Belisha, the British Transport Minister in 1934.

Biplane An aeroplane with two pairs of wings, one above the other.

Cats eyes Small glass studs protected by rubber and set into the road. They reflect the light from car headlights and show the centre of the road.

Concorde Supersonic airliner made in Britain and France which can fly at twice the speed of sound.

Dirigibles Airships (see 'Airships').

Electrical appliances Small machines or devices which are worked by electricity.

Electrification Conversion of machines or vehicles so that they work using electricity.

Festival of Britain An exhibition staged in London to celebrate British achievements. Held in 1951 – the hundredth anniversary of the highly-successful Great Exhibition of 1851.

Labour-saving device A machine that reduces the amount of human effort needed to do a job, such as a combine harvester or a dishwasher.

Laser An extremely narrow beam of light that can be used to cut materials. It is so accurate it can be used to perform delicate operations on the eye.

Mains services Household facilities such as electricity, gas and water.

Monoplane An aeroplane with one pair of wings.

Pedestrianization Turning an area, such as a street or square, into a space which only pedestrians (people on foot) can use.

Penicillin A substance first discovered by Sir Alexander Fleming. It was turned into a highly effective method of killing dangerous germs during the Second World War.

Polluting The process which spoils the air in the atmosphere (such as smoke and fumes), water in rivers or the sea, or any other aspect of our environment (such as too much noise or making an unpleasant smell).

Pylons Structures supporting electric cables.

Robots Machines that can perform tasks automatically without help from an operator.

Satellitte A small vehicle which travels around the Earth or another planet.

Supersonic Able to travel faster than the speed of sound.

Technology Using machines or mechanical means to make something work.

Telephone exchange A centre where telephone calls are connected.

Test transmission A broadcast to allow engineers to check that everything is working properly.

Transmit To broadcast or communicate.

Transplant surgery A method of replacing a diseased part of the body, such as a kidney or a heart, with a similar organ donated (given) by another patient, or taken with the consent of relatives from someone who has recently died.

X-ray The use of special rays to take photographs of the inside of a patient's body.

Books to read

(* = suitable for older readers)

Breakthrough: Communications by Philip Sauvain (Simon and Schuster, 1992)

Changing Times: Cars by Ruth Thomson (Franklin Watts, 1992)

Exploring Communications by Cliff Lines (Wayland, 1988)*

Exploring Transport by Cliff Lines (Wayland, 1988)*

How They Lived: A Teenager in the Sixties by Miriam Moss (Wayland, 1987)*

How We Used To Live, 1954–1970 by Freda Kelsall (A & C Black, 1987)*

Looking Back: Family Life by Jennifer Lines (Wayland, 1991)

Looking Back: Transport by Nigel Flynn (Wayland, 1993)

Timelines: Flight by David Salariya (Franklin Watts, 1992)

Timelines: Medicine by David Salariya (Franklin Watts, 1993)

Timelines: Transport by David Salariya (Franklin Watts, 1992)

Twenty Names in Aviation by J Hook (Wayland, 1992)

Acknowledgements
Grateful acknowledgement is given for permission to reprint copyright material:
page 4 From an essay by Dorothy Wicks in *In Those Days* edited by Julia Thorogood, copyright Age Concern Essex, Sarsen Publishing, 1994
page 5 From *The Daily Telegraph,* 22 January 1932
page 6 From *Suffolk: Within Living Memory* compiled by Suffolk Federation of Women's Institutes, Countryside Books and the Suffolk East and Suffolk West Federations of Women's Institutes, 1994
page 14 Adapted from news reports in *The Daily Express,* 6 October 1930
pages 15 and 16–17 Article 'Air Travel to the Continent' by Edward J. Hart in *Wonders of World Aviation*, The Fleetway House, c.1938
page 23 From *The Private Letters and Diaries of Sydney Moseley*, Max Parrish Publishers, 1960
page 24 Article by L. Marsland Gander in *The Daily Telegraph,* 13 May 1937
Whilst every effort has been made to trace copyright holders, the publishers apologize for any inadvertent omissions.

Index